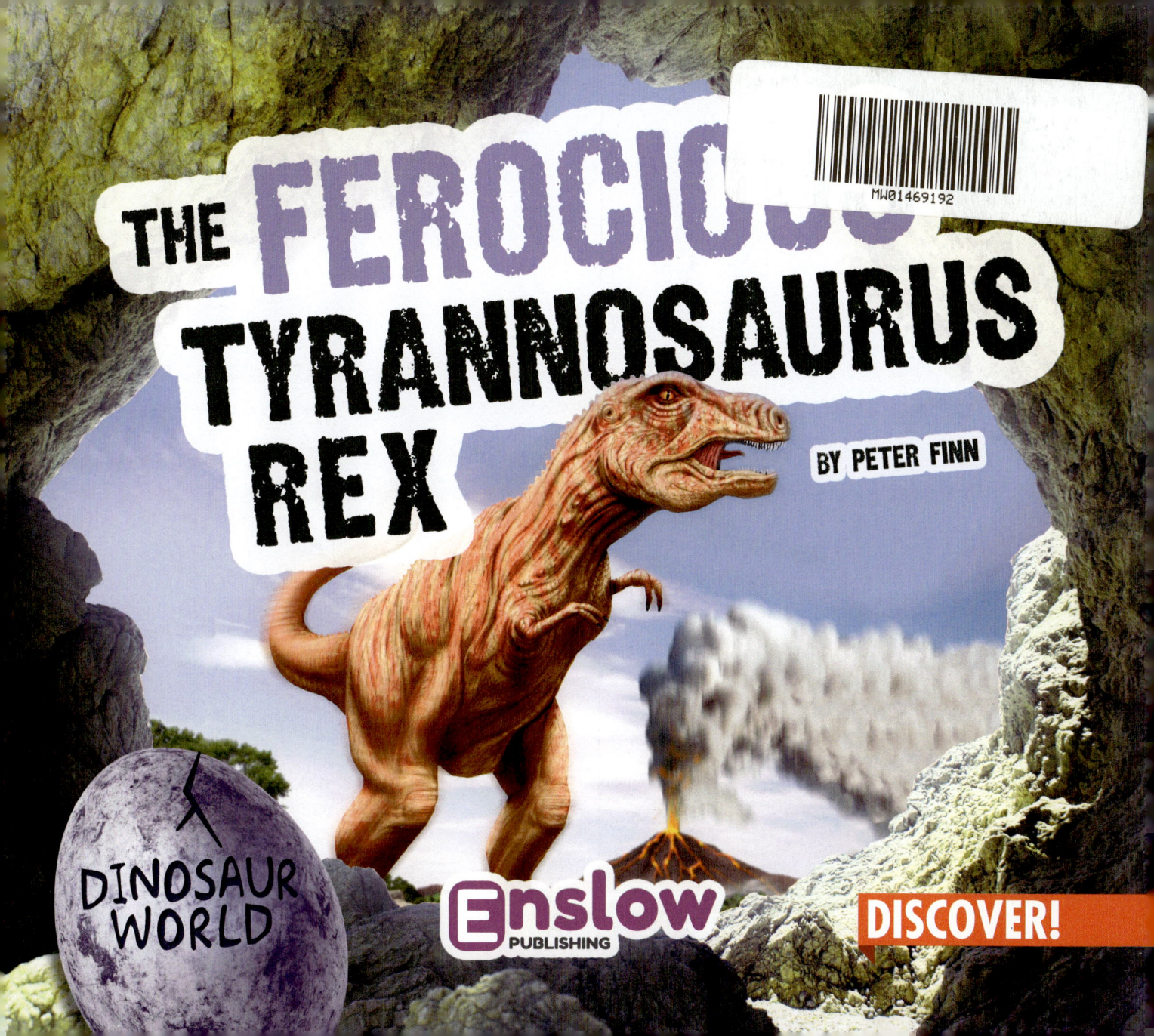

THE FEROCIOUS TYRANNOSAURUS REX

BY PETER FINN

Dinosaur World

Enslow PUBLISHING

DISCOVER!

Please visit our website, www.enslow.com. For a free color catalog of all our high-quality books, call toll free 1-800-398-2504 or fax 1-877-980-4454.

Library of Congress Cataloging-in-Publication Data

Names: Finn, Peter, 1978- author.
Title: The ferocious tyrannosaurus rex / Peter Finn.
Description: New York : Enslow Publishing, [2022] | Series: Dinosaur world | Includes index.
Identifiers: LCCN 2020048523 (print) | LCCN 2020048524 (ebook) | ISBN 9781978521049 (library binding) | ISBN 9781978521025 (paperback) | ISBN 9781978521032 (set) | ISBN 9781978521056 (ebook)
Subjects: LCSH: Tyrannosaurus rex–Juvenile literature.
Classification: LCC QE862.S3 F555 2022 (print) | LCC QE862.S3 (ebook) | DDC 567.912/9-dc23
LC record available at https://lccn.loc.gov/2020048523
LC ebook record available at https://lccn.loc.gov/2020048524

Published in 2022 by
Enslow Publishing
101 West 23rd Street, Suite #240
New York, NY 10011

Copyright © 2022 Enslow Publishing

Designer: Sarah Liddell
Interior Layout: Rachel Rising
Editor: Therese M. Shea

Illustrations by Jeffrey Mangiat
Science Consultant: Philip J. Currie, Ph.D., Professor and Canada Research Chair of Dinosaur Palaeobiology at the University of Alberta, Canada

Photo credits: Cover, pp. 1, 5, 7, 9, 11, 13, 15, 17, 19, 21 (rock border) SirinR/Shutterstock.com; pp. 2, 4, 6, 8, 10, 12, 14, 16, 18, 20, 22, 23, 24 (background) altanaka/Shutterstock.com; pp. 5, 7, 15, 19 (egg) fotoslaz/Shutterstock.com.

Portions of this work were originally authored by Daisy Allyn and published as *Tyrannosaurus Rex*. All new material this edition authored by Peter Finn.

All rights reserved. No part of this book may be reproduced in any form without permission in writing from the publisher, except by a reviewer.

Printed in the United States of America

Some of the images in this book illustrate individuals who are models. The depictions do not imply actual situations or events.

CPSIA compliance information: Batch #CSENS22: For further information contact Enslow Publishing, New York, New York, at 1-800-398-2504.

Find us on f ⓘ

CONTENTS

Meet *T. rex* 4
Long Legs 6
Short Arms 8
Big Body 10
Huge Head 12
A Feared King 14
A Scavenger? 16
Finding a *T. rex* 18
More to Learn 20
Words to Know 22
For More Information 23
Index 24

Boldface words appear in Words to Know.

MEET T. REX

Tyrannosaurus rex was a kind of dinosaur. Some call it *T. rex* for short. It lived **millions** of years ago. Its **fossils** tell us a lot about it. *T. rex* fossils have been found in the United States and Canada.

HOW TO SAY TYRANNOSAURUS REX: TIE-RAN-UH-SORE-US REX

5

LONG LEGS

Fossils show us what *T. rex* looked like. It had long, large back legs. It walked on these legs. It had a powerful tail too. Scientists aren't sure if *T. rex* was fast. Its heavy body might have made it slow.

T. REX WALKED ON ITS BACK LEGS.

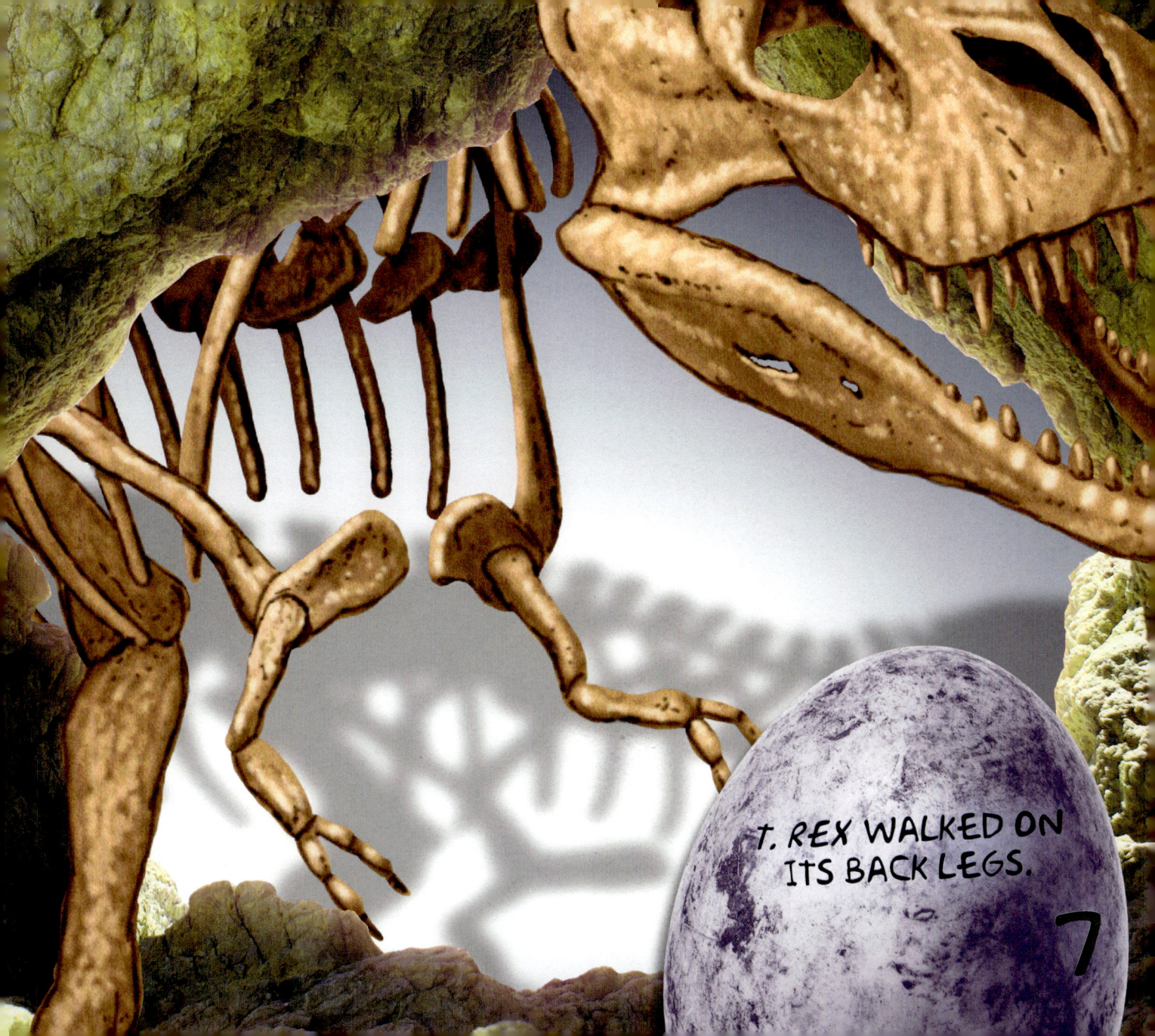

7

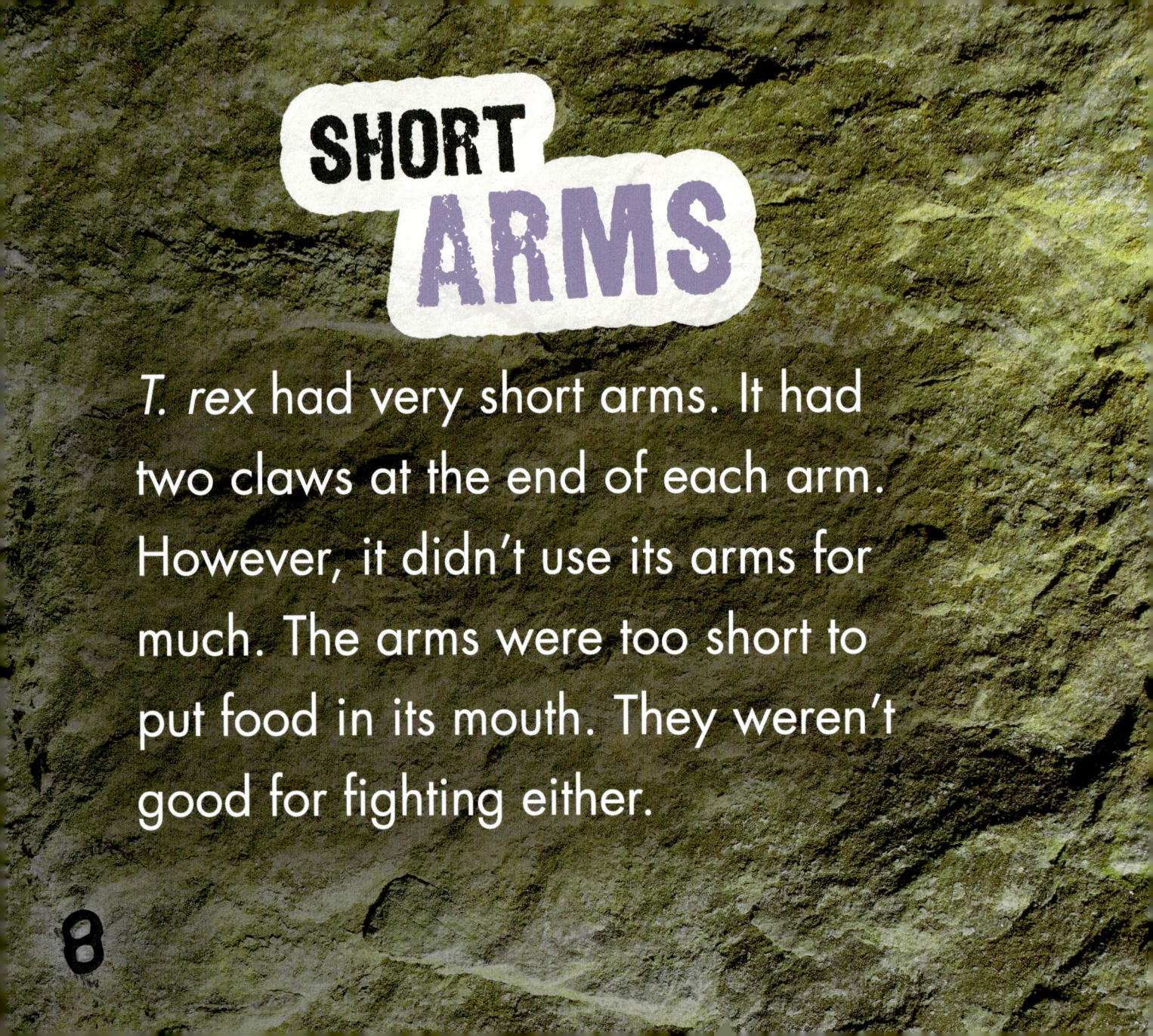

SHORT ARMS

T. rex had very short arms. It had two claws at the end of each arm. However, it didn't use its arms for much. The arms were too short to put food in its mouth. They weren't good for fighting either.

SHORT ARM

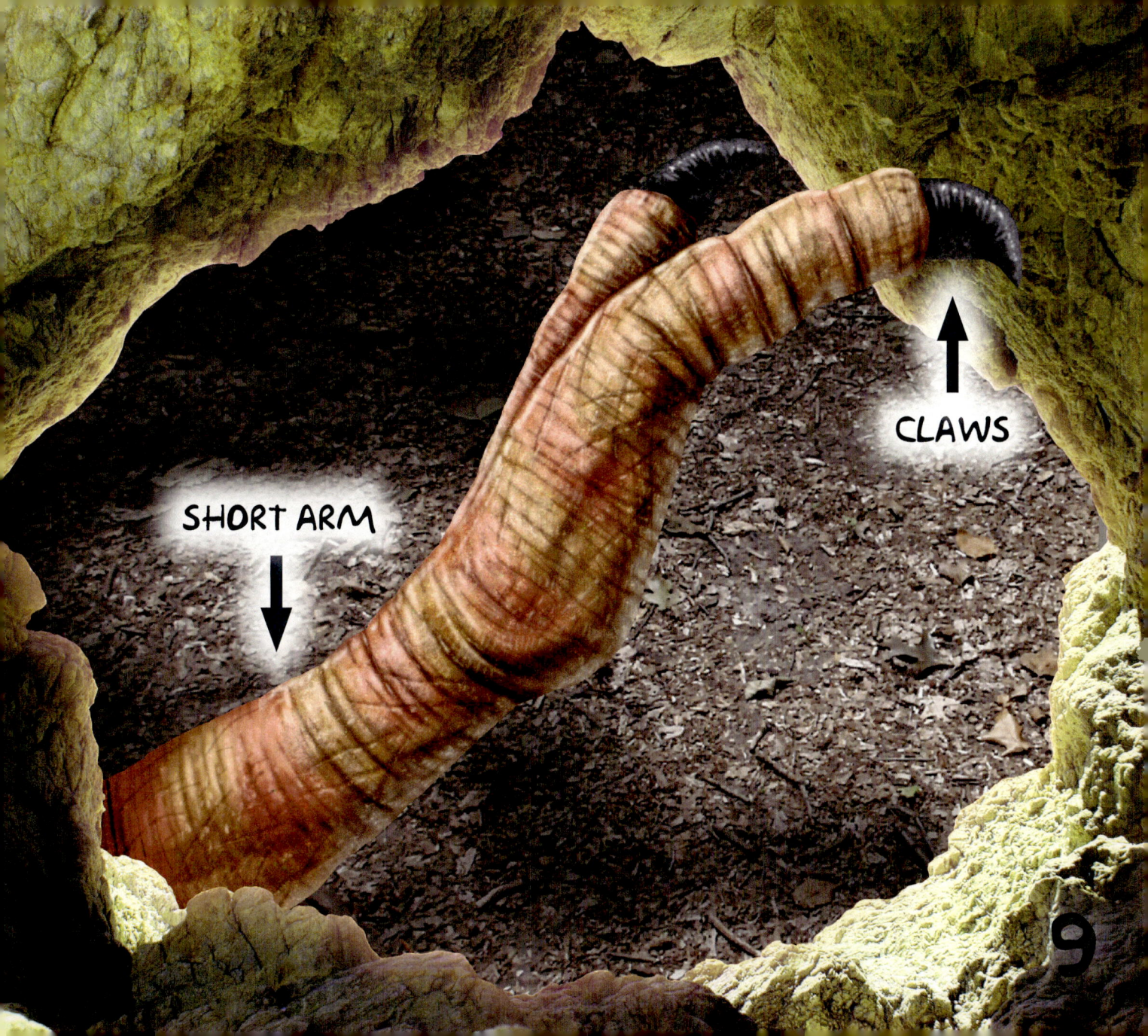

CLAWS

BIG BODY

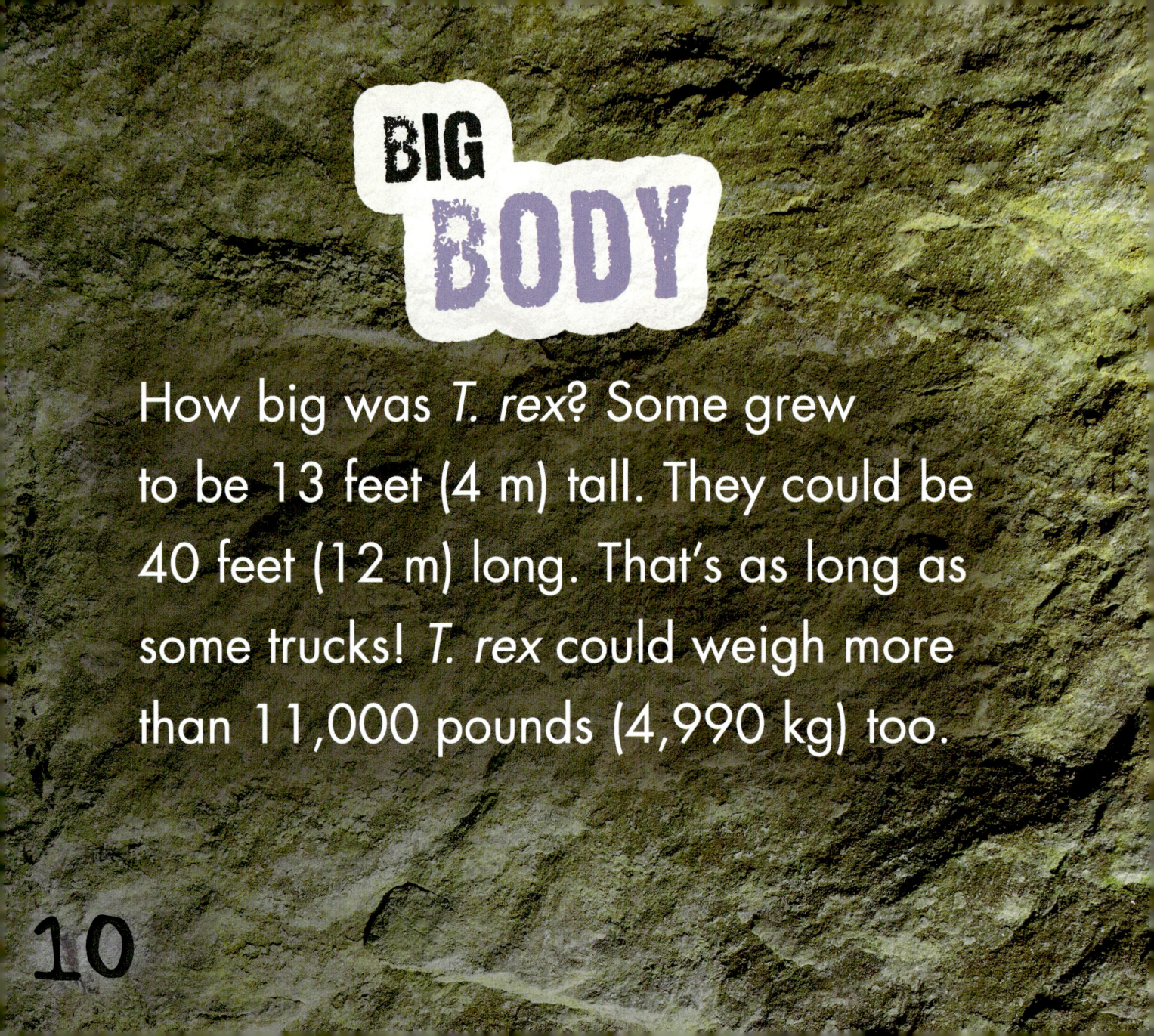

How big was *T. rex*? Some grew to be 13 feet (4 m) tall. They could be 40 feet (12 m) long. That's as long as some trucks! *T. rex* could weigh more than 11,000 pounds (4,990 kg) too.

40 FEET (12 M) LONG

HUGE HEAD

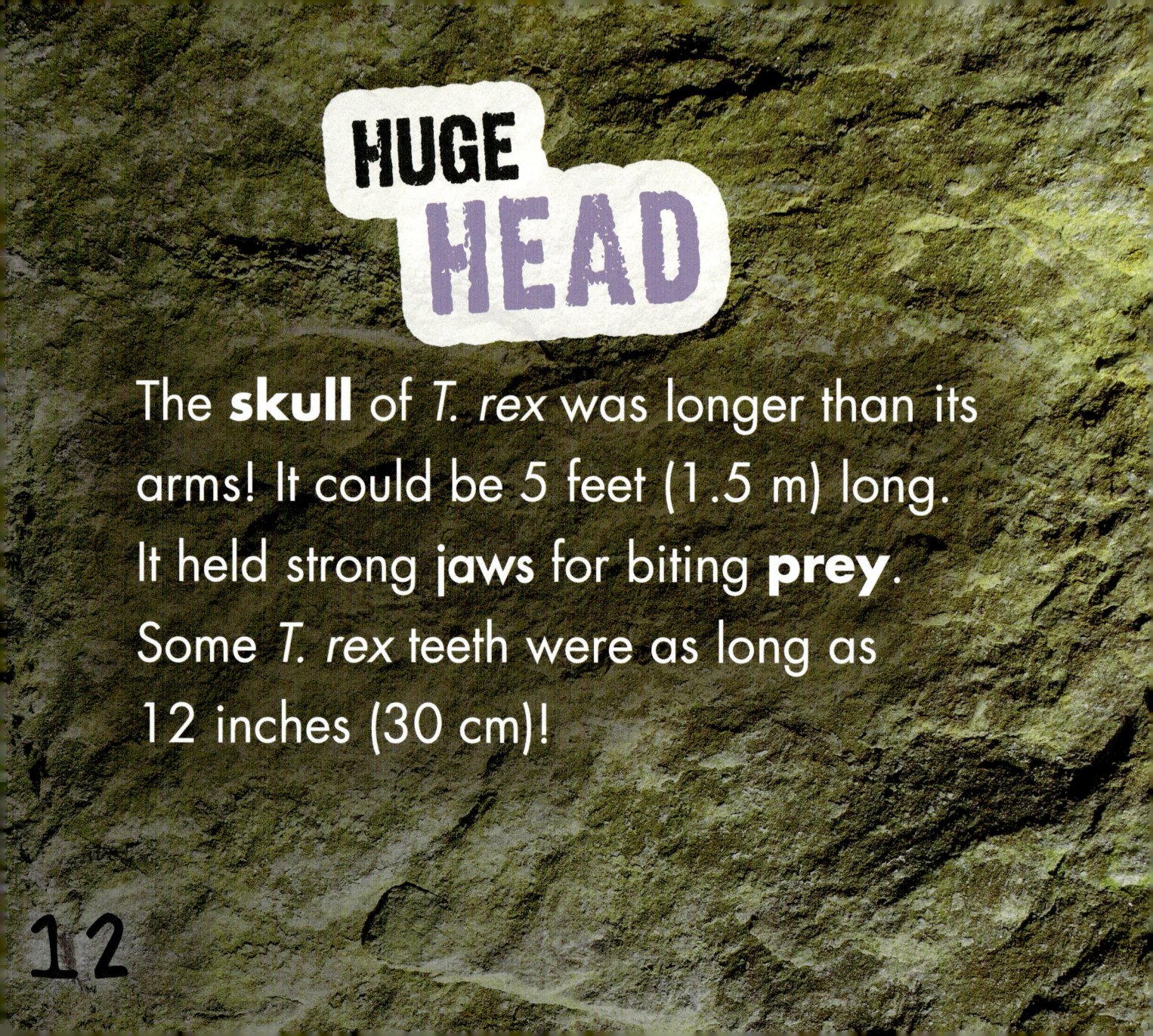

The **skull** of *T. rex* was longer than its arms! It could be 5 feet (1.5 m) long. It held strong **jaws** for biting **prey**. Some *T. rex* teeth were as long as 12 inches (30 cm)!

STRONG JAWS

LONG TEETH

A FEARED KING

Tyrannosaurus rex means "king of the tyrant lizards." A tyrant is a mean ruler that people fear. *T. rex* was a feared dinosaur. It used its powerful jaws and sharp teeth to eat other dinosaurs! *T. rex* may have hunted in packs too.

T. REX WAS A MEAT EATER.

15

A SCAVENGER?

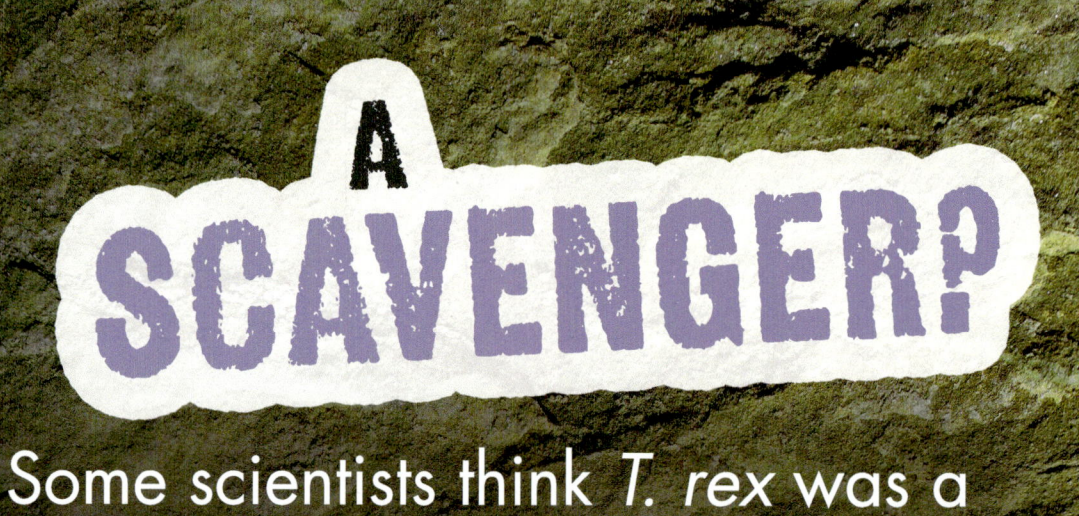

Some scientists think *T. rex* was a scavenger. Scavengers are animals that eat dead animals they find. Scientists think *T. rex* had a very good sense of smell. This sense helped *T. rex* find dead animals to eat.

GOOD SENSE OF SMELL

17

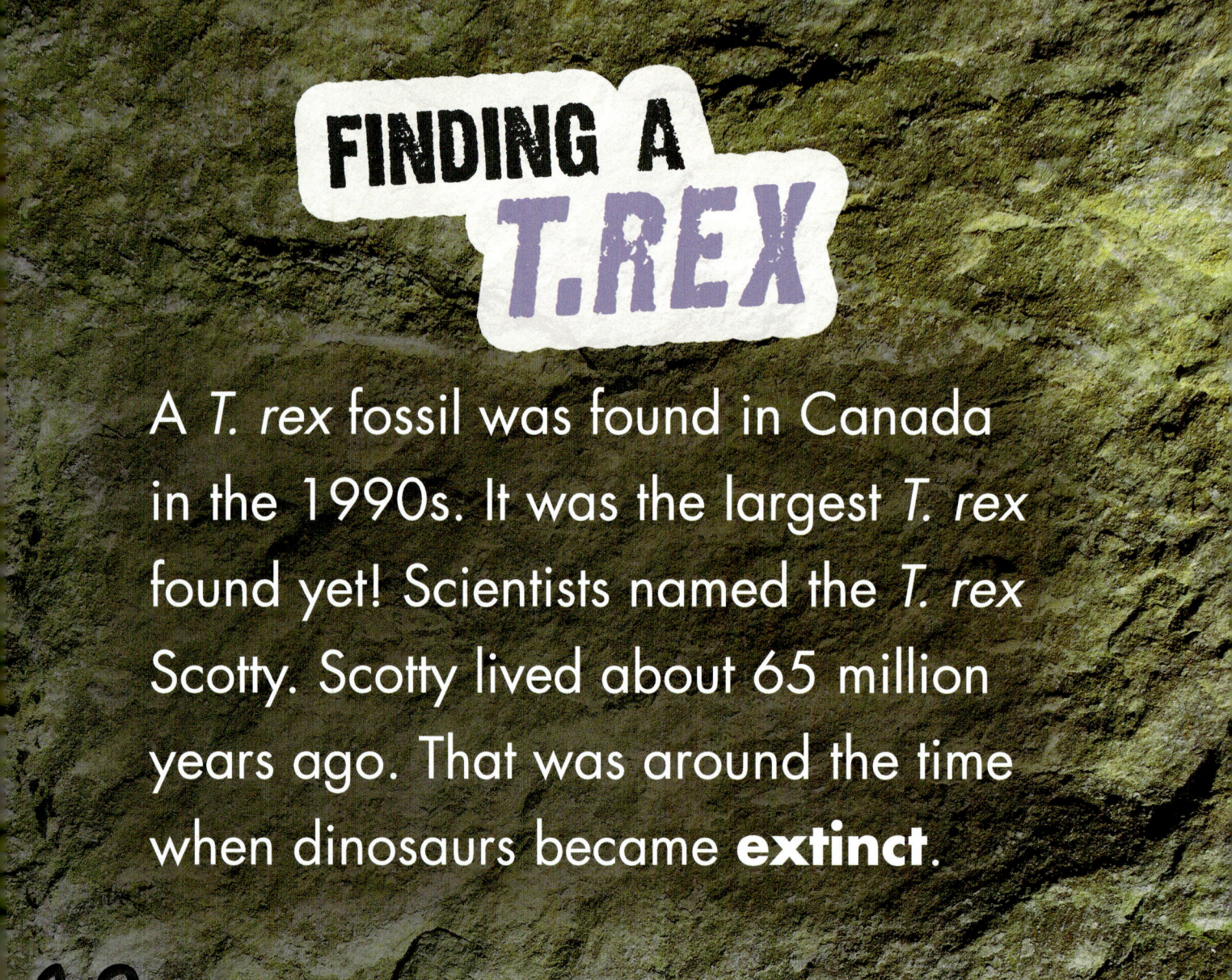

Finding a T. rex

A *T. rex* fossil was found in Canada in the 1990s. It was the largest *T. rex* found yet! Scientists named the *T. rex* Scotty. Scotty lived about 65 million years ago. That was around the time when dinosaurs became **extinct**.

T. REX BECAME EXTINCT ABOUT 65 MILLION YEARS AGO.

19

MORE TO LEARN

We have much more to learn about *T. rex*. Scientists think *T. rex* mothers laid eggs in nests like other dinosaurs did. But no *T. rex* eggs have been found yet! More fossils will teach us lots more about this dinosaur!

TYRANNOSAURUS REX

LARGE HEAD

AS BIG AS A TRUCK

STRONG JAWS

WALKED ON BACK LEGS

MEAT EATER

21

WORDS TO KNOW

extinct No longer living.

fossil The hardened marks or remains of plants and animals that formed over thousands or millions of years.

jaw The bones that hold the teeth and make up the mouth.

million A thousand thousands, or 1,000,000.

prey An animal that is hunted by other animals for food.

skull The bones that make up an animal's head and face.

FOR MORE INFORMATION

BOOKS

Allatson, Amy. *Tyrannosaurus Rex*. New York, NY: KidHaven Publishing, 2018.

Gagne, Tammy. *Tyrannosaurus Rex*. Mankato, MN: Pebble, 2019.

Waxman, Laura Hamilton. *Discovering Tyrannosaurus Rex*. Mankato, MN: Amicus, 2019.

WEBSITES

Tyrannosaurus
www.nhm.ac.uk/discover/dino-directory/tyrannosaurus.html
Read some facts about this reptile here.

What Do You Know About T. Rex?
www.amnh.org/explore/ology/paleontology/what-do-you-know-about-t.-rex
Take this T. rex quiz!

Publisher's note to educators and parents: Our editors have carefully reviewed these websites to ensure that they are suitable for students. Many websites change frequently, however, and we cannot guarantee that a site's future contents will continue to meet our high standards of quality and educational value. Be advised that students should be closely supervised whenever they access the internet.

INDEX

arms, 8, 12, 21
Canada, 4, 18
claws, 8
eggs, 20
extinct, 18, 19
fossils, 4, 6, 18, 20
jaws, 12, 14, 21
legs, 6, 7, 21
meat eater, 15, 21
nests, 20
packs, 14

prey, 12
scavenger, 16
Scotty, 18
sense of smell, 16
size, 10
skull, 12
tail, 6
teeth, 12, 14
United States, 4